On Stage

WHERE'S THE SCIENCE HERE?

On Stage

WHERE'S THE SCIENCE HERE?

VICKI COBB

Photographs by
Michael Gold

Millbrook Press   Minneapolis

To my granddaughter Lexie Cobb,
who is comfortable in the spotlight.

Acknowledgments
The author accepts full responsibility for the
accuracy of the text but gratefully acknowledges the
help of the following people: Mervyn Haines, former
Technical Director for the Public Theater of New
York, for putting me in touch with those in the know;
Greg Meeh and Patrick Boyd, of Jauchem & Meeh,
Inc., for their amazing demonstrations of special
effects; Jamie Leonard, Flying Director of "Flying by
Foy," for his expertise on the artistic defiance of
gravity; and sound engineer Stephen Manocchio, for
his cogent explanations of sound in the theater.

The photographer would like to thank the following:
Greg Meeh; Bohdan Bushell; Frederick Pisciotta;
Allison Aaron; Charles Varga; Luis Solis and the
entire staff of Jauchem & Meeh Special Effects, Inc.,
Brooklyn, NY; Professor Joseph Paparone and the
Theater Department of SUNY New Paltz, New Paltz,
NY; Stanley Rubin; Peggy Hurley; Timothy Hurley;
Theresa Cordovano; Jacqueline Cordovano; Sergio
Nazzaire; Carla Nazzaire; Michael Kurek; Nicole
Kurek; Katelyn Shoemaker; Emily Avis; and ZFX
Flying Illusions.

Cover photograph courtesy of Photofest

Interior photographs courtesy of Michael Gold with the
exception of the following: © Michal Daniel: pp. 6, 9;
© Robbie Jack/CORBIS: pp. 7, 16, 42; © Bettmann/CORBIS:
p.10; © Richard Cummins/CORBIS: p. 24; © Underwood &
Underwood/CORBIS: p. 38; Photofest: p. 43; Mark
Pennington: p. 44

Millbrook Press
A division of Lerner Publishing Group
241 First Avenue North
Minneapolis, Minnesota 55401 U.S.A.

Website address: www.lernerbooks.com

Library of Congress Cataloging-in-Publication Data
Cobb, Vicki.
On stage / by Vicki Cobb.
p. cm.—(Where's the science here?)
ISBN-13: 978-0-7613-2774-5 (lib. bdg. : alk. paper)
ISBN-10: 0-7613-2774-6 (lib. bdg. : alk. paper)
1.Optical images—Juvenile literature.   2.Visual
perception—Juvenile literature.   3. Optical
illusions—Juvenile literature.   4.Theaters—Stage-
setting and scenery—Juvenile literature.   I.Title.
QC397.5.I53C63   2005   535'.2dc22
2004029820

Manufactured in the United States of America
1 2 3 4 5 6 – DP – 11 10 09 08 07 06

Contents

Fog machines add a sense of fantasy to the theater production of *Flying Dutchman*.

It's Show Time!

The curtain goes up and it's snowing on stage. Or maybe the fog is rolling in or rain is pouring down. Two men fight and one gets thrown into a window that shatters. He lies bleeding on the ground. The scene quickly changes. We see the inside of a room with a fire in the fireplace. The shadows of tree branches are on the wall. Lightning flashes through a window. Amazingly, none of this is real! The snow is not snow, the fog is not fog, and

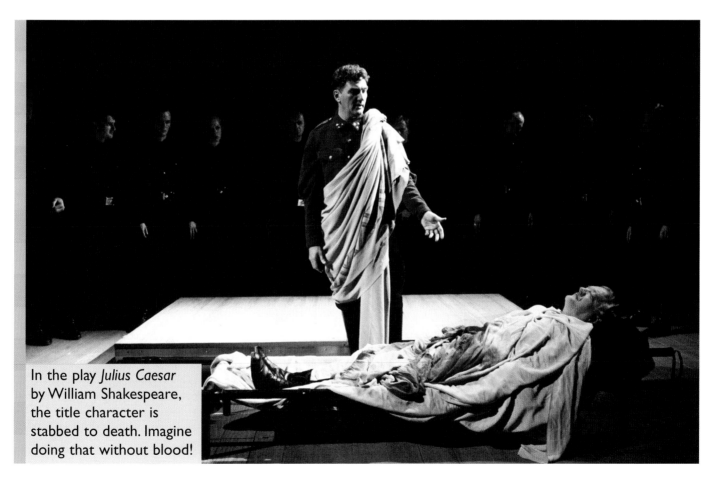

In the play *Julius Caesar* by William Shakespeare, the title character is stabbed to death. Imagine doing that without blood!

Two technicians stand in fog of their own making. This is a low fog effect that's intended to make the actors look like they're in a cloud or swamp.

the rain may be water, but there's only a 4-inch (10-centimeter) strip of it from front to back. The broken glass is not glass, and the blood is not blood. The fire, the shadows of tree branches, and the lightning are all made with lights. The magic of the theater involves special effects that create the illusion of reality for anything a playwright can dream up— including flying people and glow-in-the-dark ghosts.

Unlike movies, where spectacular special effects can be created at will by computer or camera tricks, special effects in the theater must appear and disappear at the appropriate times. They cannot be dangerous to the actors. The technicians who make a living producing special effects on stage understand the workings of rain, fog, and snow *machines*. Lighting designers know about the nature of light—how to color it, bend it, focus it, and block it. Special effects are highly technical, and there is a lot of science behind them. This book tells some of the story.

An eerie scene from *Blood Wedding* by Garcia Lorca is enhanced by a glow-in-the-dark bride rising into the air.

Fog

Real fog is made of microscopic droplets of water suspended in the air. It appears to be white because each droplet acts like a tiny mirror and reflects any light that strikes it. If you look at a beam of light, such as a car headlight, from the side when there is no fog, you will not see the beam in the air. But when a headlight shines into fog, you can clearly see the shape of the beam. The light that is reflected off such suspended particles is called the *Tyndall effect*.

Drivers find that using the high beams in fog makes it more difficult to see the road than low beams. Too much light is reflected back due to the Tyndall effect.

Another Way to See the Tyndall Effect

The Tyndall effect doesn't show up only in air. You can see it in other transparent materials, including water. Solutions in which molecules are suspended in water do not show the Tyndall effect. Salt water or sugar water are examples of such solutions. Milk, however, does not dissolve in water. Instead, tiny milk droplets are suspended evenly in the water, forming a mixture called a *colloid*.

Put a few drops of milk in a glass of water. Shine a pen flashlight beam through the glass. Look at it from the side and you will see the beam as the light reflects off the tiny, suspended milk droplets. Shine a light in salt or sugar water, and the beam is not visible.

Fog particles must be small enough to stay suspended in the air. In real fog, water molecules condense around dust particles in the air. Some special-effects fog is real. It is made when dry ice, which is solid carbon dioxide gas, is put into hot water. The heat in the water causes the solid carbon dioxide to change immediately into

This technician uses a mini smoke/fog machine to show how to make someone appear to be on fire.

a gas, which bubbles up through the water. Water vapor above the surface of the water condenses into tiny droplets around the escaping carbon dioxide gas that spills out onto the stage through a hose. This fog is heavy and stays close to the stage floor.

Most special-effects fog is made from *fog juice*—a mixture of water and a chemical called *glycol*. Glycol is a clear, syrupy liquid that freezes at a much lower temperature than water. It is used as antifreeze in car engines and is sprayed on airplane wings to de-ice them in winter. When it is heated it becomes smoky, much as burnt cooking oil is smoky. Hot glycol is used for special effects because it has no smell and it disappears quickly.

A fog machine pumps the "fog juice" through a space where it is heated. This changes the liquid into a gas, which quickly expands. When a gas expands in a closed space, the pressure increases. (You may have noticed this when a teapot whistles. Hot water vapor, or steam, builds up pressure until it comes out the hole in the teapot with enough force to create a sound.) Onstage, the hot gas is now released into the air

This individual smoke/fog machine has a reservoir for fog juice, a compressor with the heating element to make the fog, a large battery, and a switch to control the flow. It's a self-contained unit that clips onto actors, and the wires are strung under their clothing.

This smoke machine generates huge amounts of smoke. It may be used for a smoke "reveal" where a character appears or disappears in a puff of smoke.

through a nozzle. Moisture in the air condenses around the hot gas molecules, forming a fog. The advantage of this kind of fog is that it rapidly evaporates and clears the air for the next scene.

Some actors have noticed that their voices became hoarse after several performances in glycol fog. Scientists studied this problem and found out how much glycol fog was safe to breathe. The actors' union makes sure that the amount of fog is measured carefully and that it does not go above safe limits.

Make Your Own Fog

Oil droplets alone can look like fog. In fact, there are fog machines that spray mineral oil to produce theatrical fog. You can see what that looks like with a short blast of cooking-oil spray in front of a light. (Just be careful to spritz it outside or over the sink so the oil doesn't get on fabrics or hard-to-clean surfaces.)

Rain

Can you imagine *Singin' In the Rain* on stage without the rain?

Producing rain on stage is a challenge, because it must appear as if the rain is everywhere without soaking everything. The illusion of rain is produced by a *curtain* of rain-sized drops falling in front of the action. The curtain of rain is produced by "rainbars," which are 2-inch (5-centimeter) pipes suspended above the front of the stage. Two rows of nozzles, one slightly behind the other, run along the pipes. The nozzles can be adjusted so the "rain" can range from large drops to a drizzle.

Rainbars work because liquids have an amazing property when they fill an enclosed space. In this case, for example, the force of the water coming into the pipe is evenly distributed throughout the liquid. As a result, the pressure on every nozzle is exactly the same.

Onstage, the size of the raindrop depends on the size of the opening in the nozzle and the water pressure behind it. Lower pressure on a larger opening will

Hydraulic Pressure

Do a simple experiment to see how pressure travels through liquid. This kind of pressure is called *hydraulic pressure*.

Get a gallon-sized plastic bag. Poke a pushpin through both sides of the bag about ten times each, making about twenty holes. Fill the bag with water and continue holding it over the sink as you twist it closed. The water should completely fill the bag. When you squeeze the bag or poke it with your finger, you increase the pressure on the water in the bag. As a result, water squirts out the holes. Are the squirts all the same in size and strength? Poke the bag with your finger in several places. Do you always get squirts the same size? Amazingly, no matter where you push, the pressure of your fingers has the same effect on every hole.

produce large drops, while higher pressure on a smaller opening will produce a spray.

Rainbars are controlled electronically, and it is the job of the stage manager or the lighting technician to turn the rain on and off on cue. Sometimes drains are built into the stage to catch the runoff. For short rainfalls absorbent pads may catch the rain (and make it less noisy). At other times, stagehands mop up quickly between scenes.

Rainbars are horizontal pipes suspended above the stage.

The "rain" comes out of the pipe with the tiny nozzles.

Snowflakes

Real snowflakes, made of ice crystals, cannot be made in the theater without freezing the audience. So in the past, fake snowflakes were made of chopped-up chicken feathers, white-painted cornflakes, powdered potato flakes, or torn-up pieces of plastic film. But the clean-up afterward is one problem. Difficult footing for actors and dancers is another. So special-effects people had to come up with fake snowflakes that disappear when they touch a surface, just as real snow often does.

Six or eight of these snow generators (below) are used to fill a stage with snow. The snow fluid (right) is basically a soap solution similar to bubble bath.

Fake snow disappears as soon as it touches a surface.

They found that pieces of a very dry soap foam can appear like snow. A snow machine forces foam (like the foam that comes out of fire extinguishers) through a fabric-covered nozzle. A fan gently blows the foam "flakes" into the air, where they float to the stage floor.

Soap bubbles are made from an orderly arrangement of soap and water molecules. The surface of water—where it meets the air—pulls together to form a kind of "skin," called *surface tension.* Soap molecules lower the surface tension of water and make the surface skin more elastic so that bubbles can form.

A bubble itself is something like a sandwich. Soap molecules form the outer and inner surfaces of the bubble, trapping a layer of water molecules in between. The bubble's elastic skin can stretch to trap air. But the bubble is so delicate that it can be broken if it just touches another surface. After it breaks, the water quickly evaporates, leaving behind only a trace of soap.

Make Your Own Foam Flakes

Bubble-bath foam is perfect for making foam flakes. If you clap your hands on a handful of foam, you can see how realistic the flakes look and how quickly they disappear when they land.

Lighting

One of the greatest inventions of modern time is the electric lightbulb. It has transformed the way we live and the way our cities and towns appear at night. It has also transformed the theater. In the past, candles, oil lamps, and gas lamps all served one purpose—to make the stage in a dark theater bright enough so that the audience could see the action. But modern stage lighting does much more. It creates different times of day, it sets a mood, it spotlights performers, and it creates special effects, such as lightning, explosions, and slow motion.

The incandescent lightbulb produces light when electricity passes through a thin wire made of the metal tungsten. The wire becomes white hot and gives off light. The amount of light is measured in watts. Most of the bulbs in your house are 200 watts or less. The lights used on stage are much more powerful—500 to 750 watts.

Changing the color of the lighting is one way to set the mood or time of day on stage. This can be done by putting sheets of colored gels in front of the lights. Mixing colored light produces very different results from mixing colored paints. See for yourself with the following activity.

It takes a lot of lamps to give the lighting director a range of possibilities for every show.

Mixing Light

You will need three flashlights; red, green, and blue cellophane; tape or rubber bands; and some white paper. Cover each flashlight with a different color cellophane and hold the cellophane filter in place with tape or a rubber band. In a dark room, shine the colored lights on a piece of white paper. What happens when you shine the red on the green? The red on the blue? The green on the blue? What happens when you mix all three colors together?

Red, green, and blue are called *primary colors* when you mix light. Mixing light means you are adding the colors together. If you shine white light through a prism, you break it up into all the colors of the rainbow—red, orange, yellow, green, blue, and violet. The three primary light colors—red, green, and blue—contain most of the colors of the rainbow. So when you mix them together, white light is the result.

Spotlights are used to focus the attention of the audience on a particular actor or scene. A spotlight is a circle of light created by putting a lens in front of a stage lamp. The lens collects the light and focuses it into a circular area.

The lens used to make spotlights, called a *Fresnel lens*, is similar to your magnifying glass except that it is flat. The curved surface of a thick lens is put on a flat surface as a series of rings that fit inside each other. This produces a thin, lightweight lens that has the same focusing ability as a thick, heavy glass lens. Fresnel lenses are also used in overhead projectors, traffic lights, and lighthouses.

This close-up view of a Fresnel lens shows how it's made of rings of concentric circles. Notice that the angle of the outside surface of the ring closest to the center is flatter than the next ring. This angle gets progressively steeper as you move away from the center.

Focus a Light

You can see how a spotlight is focused by using a magnifying glass and a flashlight. Shine the flashlight at a wall through the magnifying glass. Move the magnifying glass back and forth. Look at the size of the bright spot in the center of the ring of light as you move the lens. When the spot is as small as you can make it, the light is focused.

Sometimes the script calls for a flash of lightning. Onstage lightning is produced by a rapidly flashing light, called a *strobe light*. A quick succession of rapid flashes, accompanied by the sound of thunder, is a realistic way of producing the effect of a thunderstorm.

Strobes can also produce a slow-motion effect. If a strobe shines on a running actor, you see his position only when the light is on him. Each flash is like a snapshot, interrupted by a very short dark period. Your mind puts these images together, and it appears as if he is running in slow motion.

Fire on stage can also be a great illusion. There are two aspects of a flame that need to be re-created in order for the audience to be convinced: light and motion. A fan blowing straight up on a piece of orange silk cut in

Limelight

In the early nineteenth century, Thomas Drummond, a British army officer, invented a lamp that created a very bright light. It worked by burning a form of limestone called *calcium carbonate*, a substance found in chalk and seashells. It was called "limelight," and it was a sharp shaft of light that could be controlled. It could light up small sections of the stage or particular actors or it could create the illusion of sunlight or moonlight. The flame used to heat the lime was produced by a torch burning hydrogen and oxygen gases—a very explosive combination. The lamp was soon replaced by the electric light. But the term "limelight" still means the bright light of special attention.

The illusion of fire is produced by movement, shiny fabric, and lighting.

the shape of a flame makes it move very realistically. Silk is a shiny material and reflects a lot of the light that strikes it. Red and orange lights reflecting off the silk make the illusion complete.

Sometimes the stage directions require a shadow.

A scene in the woods can be scarier if you see moving shadows of tree leaves on the ground. Shadows on stage are made by putting a metal *gobo* in front of the light source. Like a stencil, the gobo has shapes cut in it so it produces a pattern. Since the

The soft edges of the shadow are produced by light bending as it passes by the edge of the gobo. This kind of bending light is called *diffraction*. After the light is diffracted it spreads out, producing the fuzzy edge.

Making Shadows

Use a large desk lamp with a flexible neck to make shadows. In a dark room, place the lamp so that it shines directly on a wall several feet away. Hold your hand close to the wall. Observe the sharp edges of your shadow. Hold your hand near the lamp. Notice that your shadow is fuzzy and is darker at the center and lighter toward the edges. This happens because light rays spread apart as they travel from a source. When you hold your hand close to the wall, the rays that get past your hand don't have much of a distance to travel, so they don't spread very much and your shadow is crisp. When your hand is close to the light source, the light that is not blocked can spread, creating a fuzzy outline and a paler edge to the shadow, called a *penumbra*.

gobo is close to the light source, the shadows it produces have fuzzy, soft outlines. Sharp shadows are produced only when the light blocker is close to the surface on which the shadow falls. Sometimes you can see crisp silhouettes of the actors as they move behind a translucent screen. Then the light is behind them and they stay close to the screen.

Glow-in-the-dark paint and tape are used on the floor of the stage to mark places. When light shines on luminous chemicals in the tape or paint, it is stored as energy. When it is dark, the chemicals release the light as a soft glow. (If luminous materials are kept in the dark without being "recharged" by visible light, they will lose this glow.) This allows actors to know where to position themselves when the stage is dark. It also helps stagehands know where to put sets and props during a scene change. You may have noticed that at these times the stage is very, very dark.

Sound

Before we knew how to amplify sound electronically, actors had to project their voices so that the people in the cheap seats (far from the stage) could hear them. The *acoustics*, or the way sound behaved, in the theater could help their voices carry. But today, it is the job of the sound technician to make sure that even a stage whisper can be heard in the farthest reaches of the theater. The science of sound is crucial to a performance. Here are some of the basic ideas.

Sound is a form of energy that travels through matter—air, wood, and water are just a few of the materials that sound can travel through. A sound is caused by the vibrations of an object that sets up vibrations in the material it travels through. When these vibrations reach your ear, you hear the sound. With the human voice, the sound is generated by the vibrations of the vocal bands. You hear your own voice in two ways.

Some of it is sound that comes to your ears through the air. And some of it is sound traveling through the bones in your head. That's why your own voice sounds different to you when you hear a recording played back. Your recorded voice only travels to your ears through the air.

A microphone picks up the sound vibrations on its *diaphragm*—a tightly stretched piece of material like a drum. The vibrating diaphragm creates an electrical signal that sends the vibrations to the amplifier. The amplifier enlarges the signal and sends it to the speakers, which also have a diaphragm. The speakers change the electrical signal from the amplifier back into sound. This sound is much louder than the original sound in the microphone.

A sound technician faces many problems in the theater. That's why there are sound checks. The sound the audience hears can't be

See Sound Vibrations

Stretch some plastic wrap over the top of a large bowl like a drum. Sprinkle the surface of your "drum" with salt or sugar. Now make sounds above the drum—speak, sing, clap your hands. The plastic wrap is like a diaphragm and vibrates when sound waves reach it. The grains of salt or sugar jump around as the diaphragm vibrates.

too loud or too soft. It can't be distorted. The voice of each actor should sound like the actor and sound natural. The mike shouldn't pick up background noise, only the desired sound of a voice or music. There should be no echoes. Echoes are sound that bounces off the walls. Theaters are often designed to prevent or eliminate echoes. Usually there are no parallel walls that make sounds bounce back and forth. Often the walls are covered with a soft material that absorbs sound energy.

Actors often wear tiny microphones behind their ears or hidden in wigs that are aimed toward the mouth. Sometimes microphones are hidden in the front of the stage floor or on booms over the stage out of sight from the audience. It is important that a miked actor doesn't move in front of a speaker. This causes "feedback"—a horrible loud noise—as the sound from the mike and the speaker interact, causing the sound to amplify itself.

Sound effects add to the show. Today, sound effects are recorded on CDs and played on cue by the stage manager. Before there was recorded sound, every theater had a sound-effects person who created the sounds backstage. Thunder, for example, was created by shaking a large metal sheet, and a hoofbeat was created by striking two half-coconut shells against different surfaces in the rhythm of a horse's hooves. These kinds of simulated sound had the same kinds of sound waves as the real thing.

Broken Glass

When glass breaks, dangerous sharp edges result. Actors must be protected, so a substitute for glass is used in the theater. It must look like glass and shatter like glass, but it can't cut like glass. For many years, the most common material was hard candy—windows on stage were made of "candy glass."

Candy glass, like real candy, will dissolve in water. So it was not useful as a prop where it had to contain a liquid. A plastic called *breakaway glass* was developed for stage use. You can hold pieces of breakaway glass in your hand and squeeze. The "glass" will break into tiny shards, but you will not get cut! It is expensive and is used only as a stage prop. Apparently, inventors have found no other commercial use for this plastic.

Shards of breakaway glass look like the real thing, but you can't cut yourself on them.

A technician fills a mold with a special plastic and then removes the bottle of breakaway glass.

Making "Candy Glass"

Work with an adult, as you will be using the stove and heating sugar to a high temperature. You will need a candy thermometer, a cookie sheet, and some silicon (baker's) paper. Cover a cookie sheet with a piece of baker's paper. In a small saucepan mix 1 cup sugar with ½ cup water. Cook on a low heat, stirring constantly with a wooden spoon. The mixture will turn brown as the sugar caramelizes. When the syrup is 300°F (572°C), hard crack stage, it will be clear. Quickly pour it onto the paper-covered cookie

sheet. Tilt the sheet so the liquid spreads evenly. Leave it to cool. The paper will easily peel off the back of your golden-brown pane of "glass." Break it to see the effect. (You can also eat this "glass.") Clean the pot by soaking it in water.

Blood

A fight between characters that results in bloodshed makes a play much more realistic. Fake blood is an important material for makeup artists and actors. But colored water doesn't pass for the real thing. Blood is thicker than water, so it doesn't flow as quickly. Every fluid has a certain amount of resistance to flowing, called *viscosity*. The viscosity of blood is 3 to 4 times higher than that of water. When actors "bleed" on stage, colored water would look too fake. Ketchup is a reasonable alternative because it's thicker and flows slower. And in black-and-white movies, chocolate syrup has served as convincing-looking blood.

Cary Grant's "blood" in this old black-and-white movie is really chocolate syrup.

Actors use all kinds of tricks to make themselves appear to bleed. They can hold a capsule made of gelatin and full of fake blood in the mouth. At the proper moment, they bite the capsule and a trickle of blood appears. They can hold a bag of fake blood under the arm. A hard squeeze delivers the blood through the clothing.

Measuring Viscosity

You can compare the viscosity of different fluids that are used as fake blood. Spread newspapers or paper towels on your work space as this can be messy. Dip about 1 inch (2.5 centimeters) of the end of a clear plastic drinking straw into ketchup. Cover the open end of the straw with your finger and keep it there while you pull the straw out. (This will keep the ketchup in the straw.) Wipe the excess ketchup away from the outside of the end that was dipped. Keeping time with a stopwatch, turn the straw upside down and remove your finger. See how long it takes the ketchup to run out. Try this with syrup. Which material is more viscous? Water, of course, is so fluid that it runs out too quickly to time in this manner.

Recipe for Fake Blood

One problem caused by stage blood is keeping the costumes clean. The following recipe is supposed to wash out. However, don't try it on anything you might want to use again.

Mix a cup of corn syrup with $1/2$ cup clear liquid detergent (dishwashing soap). Stir in red food coloring until the mixture is bright red. Add a few drops of blue food coloring or some instant coffee to darken it to look more like real blood.

Paint the fake blood on a friend and see how it looks.

Sometimes an actor gets "slashed" on bare skin. "Blood" oozes out of the fake cut. This special effect is created with two-part blood, made with two clear solutions. One solution contains a kind of pesticide called *potassium thiocyanate*. The other is a chemical containing iron. Since the iron is in a compound, the solution is clear, not black like iron metal or red like iron rust.

The actor spreads one of the clear solutions on his or her skin and lets it dry. The other solution is put on the edge of a dull knife. When the two chemicals come in contact with each other, there is a reaction. The iron combines with part of the pesticide molecule to form a red, rustlike chemical that looks like blood. Since both chemicals used can be somewhat toxic, they have to be used carefully.

Coriolanus is Shakespeare's play about a conquering hero who ultimately is assassinated. It's hard to imagine losing this much real blood and still be smiling.

Flying

What could be more magical than seeing someone defy gravity and fly across the stage? You know that it's impossible, yet a part of you wishes you could be the actor and try it, just for fun. Flying across the stage is, of course, an illusion. The actor is lifted by a thin wire—so thin that it can't be seen. It is attached to a harness worn by the actor and it comes out the middle of its back. The wire is made of steel and can lift five times the weight of the person. Usually there is only one wire, but two wires attached at the hips can allow the flyer to do flips.

The flying actor is like a pendulum. This means that he or she will just swing back and forth. He must learn to hold his body in the "flying position" and to make the wire swing. Otherwise, he will hang there and perhaps twist in place.

A simple machine—the pulley—makes lifting the flyer possible. A pulley is a wheel with a groove and a rope that fits in the groove. The pulley changes the direction of the pull on the rope. A combination of pulleys keeps the rope doing the heavy lifting out of sight.

In order for stage flying to be effective, you should not be able to see the wires holding up Peter Pan.

Be a Swinger

You can experience something like what faces a flying actor by swinging on a swing in a playground. Notice that you swing back and forth under the same spot. Swings have two chains to help keep the swing straight. But an actor attached by only one wire can swing in many directions. He must learn to hold his body so that he stays straight. This takes tremendous muscle strength.

If you start swinging and then stop pumping, what happens? After a while, your swings get smaller and smaller until you stop. A flying actor must make moves to keep a flight going without looking like he's "pumping."

A series of pulleys control the raising and lowering of lights, sets, special effects, and props. There is a lot of behind-the-scenes lifting and lowering in every production. A pulley is a simple machine that changes the direction of a force and makes it easier to lift heavy objects.

The pulley just above the flyer reduces the load by half. This means that a 120-pound (54-kilogram) flyer can be lifted with 60 pounds (27 kilograms) of effort by the person pulling down on the rope.

Here's how it works: The lifting apparatus fits into a track. Once the flyer is off the ground, the fixed point over the flyer's head can be moved along the track by another backstage person. So flying takes a lot of practice by the two backstage lifters as well as by the actor that's flying.

As you can see, there is plenty of interesting work to do behind the scenes in the theater. As new technologies become available, creative minds will figure out some way to produce a special effect that makes the theater even more magical. Maybe you have some new ideas already.

Key Words

To find out more, use your favorite search engine to look up the following terms on the Internet.

acoustics	primary colors
breakaway glass	special effects
calcium carbonate	strobe light
fog juice	surface tension
Fresnel lens	Tyndall effect
glycol	viscosity

Index

Page numbers in *italics* refer to illustrations.

About the Author

Ever since *Science Experiments You Can Eat*, Vicki Cobb has delighted two generations with her scientific and playful look at the world. In the "Where's the Science Here?" series she pays attention to areas kids know are FUN, like being in the spotlight. Vicki has become quite the entertainer herself, performing for kids and their teachers in 48 of the 50 states. She also writes a weekly column for www.EducationWorld.com called "Show-Biz Science." Visit Vicki at: www.vickicobb.com.

About the Photographer

Michael Gold is a commercial photographer who has worked on assignment for some of the most exciting accounts, including *The New York Times*, *Fortune*, *Esquire*, American Express, BMW, Mobil, *Opera News*, and many more. His work includes food, internationally known celebrities, advertising, products, fashion, and corporate photography. He has had nine one-man exhibitions, portfolios published in *Popular Photography Magazine* and *Camera 35 Magazine*, and is included in "Who Needs Parks?" and *LIFE*'s first humor anthology, "LIFE Smiles Back."